Google Pixel 5 User Guide

The Complete Illustrated, Practical

Guide with Tips & Tricks to

Maximizing your Google Pixel 5

JOHN WILLIAMS

Contents

iii

A Brief Review of Google Pixel 5

Following last year's controversial Pixel 4, Google is ready to right some wrongs with the all-new Pixel 5. The Pixel 4 introduced new ideas like a telephoto camera, face unlock system, and unique gestures, but it did so at the expense of battery life. This time around, Google's trying something a bit different.

Rather than be concerned about having the highest-end processor or five cameras, Google designed the Pixel 5 to be a meat and potatoes Android phone that does everything you need it to do, has all of the core features you could ask for, and comes in at a much lower price than the latest flagships from Samsung and Apple.

How much does the Google Pixel 5 cost?

One of the most interesting things about the Pixel 5 is its price. In a world where top-of-the-line phones from most other companies cost upwards of $1,000, the Pixel 5 is refreshingly affordable at just $629. It's worth noting that there's just one storage configuration being offered, with the Pixel 5 coming with 128GB of space for your apps, games, and movies.

When you compare this side-by-side with 2019's Pixel 4 series, it's pretty impressive how much more affordable the Pixel 5 actually is. The 128GB Pixel 4 retailed for $899, with the 128GB Pixel 4 XL costing $999.

There are some spec changes/cuts that allowed Google to hit that lower price tag, but especially in the economic climate we live in today, we're more than happy to see Google focusing on affordability for its top offering.

What are the specs for the Google Pixel 5?

Display	6-inch OLED

	Full HD+ 90Hz refresh rate 19.5:9 aspect ratio Gorilla Glass 6
Operating System	Android 11
Processor	Qualcomm Snapdragon 765G
Rear Camera 1	12.2MP primary camera f/1.7 aperture OIS
Rear Camera 2	16MP ultra-wide camera f/2.2 aperture
Memory	8GB of RAM
Storage	128GB
Battery	4,080 mAh
Charging	18W wired charging

15W wireless   charging

5W   reverse   wireless

charging

Security   Fingerprint sensor

Okay, so the Pixel 5 has an amazing price. But how in the world did Google pull it off? A quick look at the spec sheet makes it clear why the Pixel 5 is able to cost as little as it does.

The main takeaway here is that Qualcomm Snapdragon 765G processor. All previous Pixels (not counting Pixel a devices) have shipped with a Qualcomm Snapdragon 800-series chip — the highest-end option offered by Qualcomm. Other flagships like the Galaxy S20 and OnePlus 8 Pro utilize the Snapdragon 865, but that's not the case for the Pixel 5.

The Snapdragon 765 chipset inside the Pixel 5 is more than powerful enough for any app or game you throw at it, and with this decision, you get to benefit from high-end performance without having to pay out the nose for it.

Looking at some of the other specs, it's great to see that Google once again offers a 90Hz refresh rate for the Pixel 5's display, along with an OLED panel for great colors and deep blacks. The 4,080 mAh battery is also *substantially* larger than the 2,800 mAh unit the Pixel 4 shipped with, which should allow for a night and day difference when it comes to battery life. All of the makings are here for the Pixel 5 to be one of the new best Android phones, and when you factor in that low price tag, it becomes that much more impressive.

Does the Pixel 5 still have the best camera?

Camera quality is always the number one reason to buy a Pixel phone over anything else

— that's why the Pixel 4 XL, despite its problems was one of the best Android phones you can buy — and for the Pixel 5, Google's trying something it's never done before.

The primary camera is the exact same 12.2MP sensor that we've had for years, but on the Pixel 5, it's joined by a 16MP ultra-wide camera. This is the first time any Pixel phone has shipped with an ultra-wide lens, and the telephoto one that was introduced on the Pixel 4 is nowhere to be seen.

Google made a big deal last year about how telephoto was more important than ultra-wide, so it's a bit funny to see the company correcting course one generation later. The telephoto camera on the Pixel 4 was very good, but the switch to ultra-wide is one we're excited to see.

Outside of that, you'll notice those are the only cameras present on the back of the Pixel 5. In a world where so many phones are shipping with four or five different rear cameras, Google's decision to only offer two does go against market trends. However, as we've seen time and time again, having one or two excellent cameras instead of a bunch of extra ones that aren't any good is usually preferable.

Is there a Google Pixel 5 XL?

Speaking of changes to the Pixel family, you might have noticed by now that we haven't said anything about the Pixel 5 XL. That's because there isn't one ☐. While we've typically gotten a Pixel 4 and 4 XL, Pixel 3 and 3 XL, and so on, the regular Pixel 5 is the only entry in the entire Pixel 5 family.

As Alex Dobie explained this past September, Google seems to have gone through a few different branding decisions while creating this year's Pixels. The phone that we have now as the Pixel 4a 5G is thought to have originally been the Pixel 5, with the Pixel 5 we're talking about today previously referred to as the Pixel 5s — a higher-end, more premium version of what would have been the Pixel 5 but is now the Pixel 4a 5G.

You could argue that the Pixel 4a 5G is a pseudo-Pixel 5 XL given its slightly larger 6.2-inch display, but that phone sacrifices things like the 90Hz refresh rate, wireless charging, and has slightly less RAM.

How to setup your device

When customers first turn on their new devices, they may see a setup wizard to help get started. Take a look at the steps below to help walk customers through the setup process if needed.

- Insert your SIM card, then press the Power button to turn your device on.

- At the welcome screen, tap Start.

- If you haven't inserted your SIM card, you'll be prompted to insert it now. Don't tap SIM-free setup, it's not supported by T-Mobile.

- Tap your Wi-Fi network to enter the password and connect. Tap Skip to move forward without connecting to Wi-Fi.

- Wait while the device checks for and installs any available updates.

- On the Copy apps & data screen, tap Next to copy data from another device.

- Tap Don't copy to skip the data copy and set up as a new device.

- Follow the on-screen prompts to begin copying your previous data to the new device.

- Sign in or create your new Gmail account. Once you have signed into your account, any backed up data will begin to download.

- Review the Google Terms of Service. If you agree to them, tap I agree.

- Review the available Google services and tap the switches to turn them on or off. Make sure to scroll to the bottom, then tap Accept.

- Review the additional legal terms, then tap I accept to accept.

- Tap Next to set up fingerprint unlocking with Pixel Imprint, then follow the steps to set up your PIN, Password, and Fingerprint. Otherwise, tap Skip.

- If you didn't set up your Pixel Imprint, you'll be prompted to set up your screen lock. Follow the on-screen step to set it up, or tap Skip.

- Tap Continue and follow the on-screen steps to set up additional services like Google Assistant, Google Pay, and your Wallpaper. Otherwise tap Leave & get reminder.

- Review any additional apps, check the ones you want to install, then tap OK.

- Make sure you install the T-Mobile apps, like Visual Voicemail and Device Unlock.

- If you don't install an app now, you can use the Play store to install it laterr.

- Review and agree to T-Mobile's Device Data Collection. When done, your device will be ready to use.

How to Backup Contacts to Google

Don't lose your contacts when you get a new phone. If you own an Android phone, Google will automatically backup your contacts, app data, call history, and more to Google Drive. This is turned on by default. When you log into your Google account on your new phone, it should sync your data automatically. But, if you want to manually synch your contacts or export them into a separate file, then upload them to Google yourself, here's how.

How to Sync Contacts to Google Account Manually

Manually syncing refreshes your account data for all Google apps, including ones that have auto-sync turned off.

- Open your smartphone's Settings.

- Tap Accounts.

- If you have multiple accounts, tap the one you want to update.

- Tap Account sync > More (the three vertical dots) > Sync now.

How to Backup Contacts on Android By Exporting Them

You can take the contacts stored on your device's hard drive or SIM card and export them into a .vcf file. This will let you import those contacts to a new phone.

- Open the Contacts app on your Android phone.

- Tap Menu > Settings > Export.

- If you have multiple accounts, choose which one you want to export contacts from.

- Tap Save to download the .vcf file.

- Once you have that file, you'll need to store it in a safe place, either on

13

removable storage like a SIM or memory card, or in the cloud via services like Google Drive or Gmail.

How to Import Contacts From a .VCF File

To upload your saved .vcf file to a new phone:

- Open the Contacts app and tap Settings > Import > .vcf file.

- In the Downloads manager, tap the Menu icon and navigate to where you saved the file (Google Drive, a SD card, etc.).

- Once you tap the .vcf file, Google will automatically import the data to your phone.

How to switch from iOS to android

While the Android OS and Apple's iOS each have fiercely loyal users who would never imagine switching to the other platform, it does happen. In fact, many people switch more than once before choosing a winner. An Android user might get fed up with the operating system's fragmentation or an Apple user may tire of the walled garden and take the plunge.

With that switch comes a learning curve and the daunting task of transferring important data, including contacts and photos and setting up apps. Switching from iOS to Android doesn't have to be difficult, especially since many Google-centric apps are available on iOS, making it easier to back up certain data. Just be prepared to spend some time getting used to the new interface.

The directions below should apply no matter who made your Android phone: Samsung, Google, Huawei, Xiaomi, etc.

- **Set Up Gmail and Sync Contacts**

The first thing you need to do when you set up an Android smartphone is to set up a Gmail account or log into it if you already use it. Aside from email, your Gmail address serves as a login for all Google services, including the Google Play Store. If you already use Gmail and have synced your contacts to it, then you can simply log in and your contacts will transfer to your new device. You can also transfer your contacts from iCloud by exporting them as a vCard and then importing them into Gmail; you can also sync your contacts from iTunes.

Not sure where your contacts are saved? Go into settings, then contacts, and tap default account to see which is selected. Finally, you can import your contacts using your SIM card or a third-party app, such as **Copy My Data**, **Phone Copier**, or **SHAREit**.

Google Drive for iOS now has a feature that lets you back up your contacts, calendar, and camera roll. It may take a few hours the first time you do it, but it'll save a lot of time when you switch over to Android.

If you have email on other platforms, such as Yahoo or Outlook, you can set up those accounts, too, using the Android Email app.Next, you'll want to sync your calendar with Gmail, if you haven't already, so you don't lose any appointments. You can do this easily in your iPhone settings. Google Calendar is also compatible with iOS devices, so you can still coordinate with other iOS users and access your calendar on an iPad.

- **Backing Up Your Photos**

The easiest way to move your photos from your iPhone to Android is to download the Google Photos app for iOS, sign in with your Gmail, and selecting the back up & sync option from the menu. Then download Google Photos on your Android, sign in, and you're done. You can also use a third-party app, such as **Send Anywhere**, or your preferred cloud storage software, such as **Dropbox** or **Google Drive**.

- **Transferring Your Music**

You can also move your music using cloud storage or you can transfer up to 50,000 of the songs from your iTunes library to Google Play Music for free. Then you can access your music from any web browser and all of your Android devices. First, make sure your iPhone or iPad is synced with iTunes, then install the Google Play Music Manager on your computer, which will upload your iTunes music to the cloud. Even though Google Play Music is free, you'll have to set up payment information for future purchases.

Alternatively, you can import your music into another service such as Spotify or Amazon Prime Music. In any case, it's always a good idea to regularly back up your music and other digital data.

- **Bye Bye iMessage**

If you've been using iMessage to communicate with friends and family, you'll have to find a replacement as it's not available on Android devices. Before you get rid of your iPhone or iPad, be sure to turn it off so that your messages don't continue getting redirected there, for example, if another iOS user texts you using your email address. Just go into settings, select messages, and turn IMessage off. If you've already ditched your iPhone, you can contact Apple and ask them to deregister your phone number with iMessage.

Tips to Navigation

While the Pixel 4 doesn't debut Android Pie - it's also on older Pixels and some other devices - it does introduce a fundamental change in navigation around Android, so we'll get you started there.

How to access Overview

Overview replaces one of the the functions of the recent apps button. A short swipe up will see the UI pop into Overview. This gives you cards for your apps which you can swipe away to the top to close, or scroll left and right through.

Quickly switch apps

Previously, a double tap on the recent apps button would switch between the last app and the current. Now that's replaced by a swipe right on the home button. For a longer scroll, you can drag to the right and hold and you'll enter a carousel to scan left or right to apps. It's basically the same as accessing Overview and scanning left or right, but it can be done with one press.

Close all open apps view overview

To shut all your apps down, you can either swipe them all away to the top in Overview, or you can scroll all the way to the end of the list and tap "clear all". That will clear out all your recent items.

How to launch Google Assistant

As before - and on most Android devices - a press and hold of the home button will launch Google Assistant. The Pixel 3 devices will also let you squeeze the handset to launch Google Assistant. You can find the options in settings > system > gestures > active edge.

How to open the apps tray

Yes, it's still a swipe up from the bottom of the display, but with Overview in the mix, you'll need a longer swipe to bypass Overview and head straight into the apps. It does work, it just takes a little getting used to.

Enable suggestions on Overview

Android Pie has "suggestions" in a couple of places. These suggestions come from your app use, so it can suggest apps you might be trying to access quickly. These have been available at the top of the apps tray for a few years, but now come to Overview - so in many cases, opening the apps tray isn't necessary. You can find the option in home settings > suggestions.

Night Sight Mode

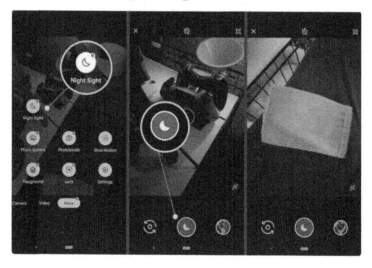

Lighting vexes the best photographers, but this function changes all that. By choosing the Night Sight option, you enable the Pixel's HDR+ processing to boost colors and brightness. If the camera detects a dark environment, a pop-up suggestion automatically appears. Night Sight is best used in rooms with dim lighting or evening shots you want to appear to look day-lit.

Night Sight is yet another example of how artificial intelligence (AI) is revolutionizing the photography world. Fortunately, you don't have to understand how it works to use it.

Which Phones Have Night Sight

All Pixel phones have this function, but they don't all work the same way. On Pixel 1 and 2, a modified HDR+'s merging algorithm is used to help detect and reject misaligned pieces of frames.

On Pixel 3, 4 or 5, a similarly re-tuned Super Res Zoom is used whether you zoom or not. Although it was developed for super-resolution, it also works to reduce noise, since it averages multiple images together. Super Res Zoom produces better results for some nighttime scenes than HDR+, but requires the faster processor of the Pixel 3, 4 or 5.

How Does Night Sight Work?

Night Sight is designed to take better photos in low-light conditions for both the rear and front-facing cameras. It allows you capture vibrant and detailed low-light photos without the need for a distorting flash or tripod. And, like night goggles, it'll even work in light so dim you can't see much with your own eyes.

Shooting in low light can be infuriating for even the best photographers. Google has tapped into its bodacious Pixel HDR+ algorithm to boost color, brightness, and stability when confronted by low light. By choosing the Night Sight option, you enable Pixel's HDR+ processing to boost colors and brightness. If the camera detects a dark environment, a pop-up suggestion automatically appears.

It's All About the HDR+

Google's HDR+ Processing is a proprietary technology that reduces "noise" and enlivens colors. In reality, it's taking a burst of shots, then combining the best of each image to create one final version of that image.

Oddly enough, HDR+ isn't enabled by default; you need to go into the camera's advanced settings and enable the HDR+ control. Since the Pixel's HDR+ is the best and most important part of its picture taking process, it's essential you turn this on the minute you buy your phone.

Night Sight is constantly adapting to both you and your photo object. As you press the shutter button, Night Sight measures for any hand shake, as well as any motion in the scene and then compensates by using using shorter exposure bursts.

If stability isn't an issue, Night Sight focuses its processing power on capturing light to brighten the scene. It takes multiple photos, then, by merging these multiple exposures, prevents motion blur and brightens the photo, resulting in a well-lit and sharp photo.

Some critics have accused Night Sight of fabricating a photo — taking some basic visual data and then filling in the blanks with educated guesses — and they're not entirely off-base. Night Sight is essentially an improvement of a photo technology called image stacking, which has been around for years. And yet, Night Sight is turning heads even among SLR camera buffs.

How to Use Night Sight

Night Sight is automatically enabled on your device, and there are two ways of accessing this feature:

Automatically: If you're taking a photo in low light, Pixel will suggest using Night Sight. The small button will appear on the screen and you need only tap to initiate this function.

Selecting Night Sight manually

Manually: If the Night Sight option is not automatically triggered, but you want to brighten the shot, you can select Night Sight mode by clicking the MORE button, just to the right of VIDEO mode.

How to Master Night Sight

Google has listed some tips to help users take full advantage of Night Sight mode. Some of its suggestions include:

- **Motion**: Ask your photo subject to hold still for a few seconds before and after you press the shutter button.

- **Stability**: Prop the phone against a stable surface, if possible. The steadier the hand, the more processing can focus on the light and sharpness of the exposure.

- **Focus**: Tap both on and around your subject before taking the picture. This step helps your camera focus when taking photos in very dark conditions.

- **Bright Lights**: Avoid any bright light sources in a Night Sight picture. It'll cause unwanted reflections in your photo.

How to pick a live wallpaper

The Pixel offers a range of "live" wallpapers, with subtle active elements in them giving some movement to your home screen. Long press on the home screen and select wallpaper. Then head to the "living universe" section and you'll find those live wallpapers.

Turn printing on or off for your phone

1. Open your phone's Settings app.

2. Tap **connected devices** › **Connection preferences** › **printing**.

3. Tap a print service, like **Cloud Print**.

4. Turn the print service on or off.

Print from apps

How to print depends on the app you're printing from. Often, you can tap Menu ≡ or More ⋮ , and then tap **Print**.

Not all apps work with printing. In apps that can't print, you can take a photo of the app's screen and then print the photo

Print Gmail messages

You can print individual messages or all messages inside a conversation.

Note: Make sure you've connected your printer using Cloud print or Air print

Use Google Cloud Print

To see and choose from available printers:

1. Open your phone's Settings app.

2. Tap **connected devices** ⟩ **Connection preferences** ⟩ **printing**.

3. Tap **Cloud Print**.

Add & use a new print service

To add a new print service:

1. Open your phone's Settings app.

2. Tap **connected devices** › **Connection preferences** › **printing**.

3. Tap **Add service**.

4. Enter the printer information.

To use a new print service after adding it:

1. Open your phone's Settings app.

2. Tap **connected devices** › **Connection preferences** › **printing**.

3. Tap the print service, like HP.

Print a single email

If there are multiple emails within the same conversation, you can print just one of those emails.

1. On your computer, go to Gmail.

2. Open the email you want to print.

3. In the top right of the email, click More

4. Click **Print**.

Print an email with replies

If there are multiple emails within the same conversation, you can print all those emails together.

1. On your computer, go to Gmail.

2. Open the conversation you want to print.

3. In the top right, click Print all🖶

Print your calendar

If you ever need a paper copy of your calendar, you can print one by week, month, or any date range you choose.

1. On your computer, open Google Calendar.

2. In the top right, click **Day**, **Week**, **Month**, **Year**, **Schedule**, or **4 Days** to choose which date range to print.

3. In the top right, click Settings ⚙ › **Print**.

4. On the Print Preview page, you can change details like font size and color settings.

5. Click **Print**. A window with print options will appear.

6. At the top left, click **Print**.

Print a document

1. On your computer, open a document in Google Docs.

2. Click **File** > **Print**.

3. In the window that opens, choose your print settings.

4. Click **Print**.

Print a spreadsheet

1. On your computer, open a spreadsheet in Google Sheets.

2. Click **File** > **Print**.

3. Optional: Choose your print settings, like margins or page orientation.

4. Click **Next**.

5. In the window that opens, choose your print settings.

6. Click **Print**.

Print a presentation

1. On your computer, open a presentation in Google Slides

2. Click **File**.

   o **Print without changes**: Click **Print**.

   o **Adjust orientation**: Click **Print settings and preview** ˃ in the toolbar, click **Handout** ˃ **Landscape.**

   o **Print with speaker notes**: Click **Print settings and preview** ˃ in the   toolbar,   click   **1   slide with notes**.

3. Click **Print**.

4. In the window that opens, choose your print settings.

5. Click **Print**.

Print your file as a PDF

1. On your computer, open the document, presentation, or spreadsheet you want to print.

2. Click **File** ˃ **Print**.

   ○ **Document or Presentation**: A PDF   file will   automatically download.

   ○ **Spreadsheet**: In the window that opens, choose your print settings. ˃ Click **Next**. A PDF file will automatically download.

3. When the download completes, open the file.

4. In your PDF viewer, go to **File** ˃ **Print**.

5. In the window that opens, choose your print settings.

6. Click **Print**.

Print your file in another file format

1. On your computer, open the document, spreadsheet, or presentation you want to print.

2. Click **File** ˃ **Download as**.

3. Select the file format you want.

4. Find the downloaded file on your computer, and open it.

5. Print the downloaded file.

Change page setup of a Google Doc

1. On your computer, open a document in Google Docs.

2. In the toolbar, click **File** › **Page setup**.

3. Go to the setting you want to change:

   ○ Orientation

   ○ Paper size

   ○ Page color

   ○ Margins

4. Make your changes.

5. Click **OK**.

6. Optional: To make new documents open with the settings you chose, click **set as default.**

Google Sheets: Change the orientation, paper size, and margins when you print a spreadsheet.

Google Slides: Change the size of your slides.

Get help in an emergency using your Pixel Phone?

You can use your Pixel phone to save and share your emergency info. In some countries and with some carriers, your phone can contact emergency services automatically.

Add emergency info to your Pixel 5

You can add a link to personal emergency info to your phone's lock screen. For example, you can add info that would help first responders in an emergency, like your blood type, allergies, and medications.

Important: Anyone who picks up your phone can view your message and emergency info without unlocking your phone.

1. On your Pixel 5, open the Safety app .

2. If asked, sign in to your Google Account.

3. To add medical info: Tap Settings ⚙ ›
   Medical information.

- ○ To enter info like your allergies or medications, tap an item in the list.

4. To add an emergency contact: Tap Settings ⚙ › **Emergency contacts** › **Add contact** › one of your contacts.

Tips:

- To show your emergency info when your screen is locked, tap **Show when device is locked** › **Show when locked**.

- Set up a SIM card or eSIM with your phone. Otherwise, your phone can't text your emergency contact later.

Turn on car crash detection

If your Pixel 5 determines you've been in a car crash, it can help call emergency services automatically, like 911 in the US, and share your location. Car crash detection is available in the United States and in English only.

1. If you haven't yet, add a SIM to your Pixel 5. Open the Safety app .

2. Tap Settings⚙.

3. Under "Driving," tap **Car crash detection**.

4. Turn on **Car crash detection**.

5. When asked to share your location, tap **Allow all the time**.

6. When asked to share your microphone and physical activity, tap **Allow**.

Put a message on your lock screen

You can add a line of text to your lock screen, like info that would help someone return your phone if you lost it.

1. Open your phone's Settings app.

2. Tap **Display** ⟩ **Advanced** ⟩ **Lock screen display** ⟩ **Lock screen message**.

3. Enter your message.

4. Tap **Save**.

Control emergency broadcast notifications

You can change your emergency alert settings, like for AMBER alerts and threat notifications.

1. Open your phone's Settings app.

2. Tap **Apps & notifications** ⟩ **Advanced** ⟩ **Emergency alerts**.

Get help after a car crash

Important: Car crash detection doesn't work in Airplane mode, or when Battery Saver is on.

If your phone determines you got in a car crash, and if you turned on car crash detection previously:

1. Your phone will vibrate, ring loudly, and ask if you need help, both aloud and on your phone screen.

2. Respond within 60 seconds:

- ○ **To call emergency services:** Say "Emergency" or tap the emergency button twice. Your phone will turn on speakerphone automatically **to not call:** Say "Cancel" or tap **I am ok**. Your phone won't make an emergency call.

- ○ **If you don't respond:** Your phone will automatically turn on speakerphone, try to call emergency services, say that a car crash happened, and share your device's approximate location. The message will repeat, but you can speak over it. To stop the message and stay on the call, tap **Cancel**.

Text your emergency contact

1. On your Pixel phone, open the Safety

2. Tap **Start message**.

3. Enter your message.

4. Choose which emergency contact you want to message.

5. Tap **Send**. Your phone will send your contact an SMS text message with your message and location.

Find emergency info

1. On a locked screen, swipe up.

2. Tap **Emergency** › **Emergency information**.

3. When **Emergency information** flashes, tap it again.

Send your location automatically

To help first responders find you quickly, dial an emergency number. For example, dial:

- 911 in the U.S.

- 112 in Europe

If Android Emergency Location Service (ELS) works in your country and on your mobile network, and you haven't turned off ELS, your phone will automatically send its location using ELS.

Turn Emergency Location Service on or off

You can turn emergency location services on or off at any time.

1. Open your device's Settings app.

2. Tap **Location** › **Advanced** › **Emergency Location Service**.

3. Turn **Emergency Location Service** on or off.

How car crash detection works

Your Pixel 5 can use information such as your phone's location, motion sensors, and nearby sounds to detect a possible car crash. Car crash detection requires location, physical activity, and microphone permissions to work

Your Pixel phone might not be able to call emergency services in some cases, including when your phone is connected to a weak or unreliable mobile network, in an ongoing call, or in an area with less reliable mobile networks. Your phone may not be able to detect all crashes. High impact activities could trigger calls to emergency services.

How Emergency Location Service works

Your phone uses ELS only when you call or text an emergency number.

During your emergency call, ELS may use Google Location Services and other information to obtain the most accurate location possible for the device.

Your phone sends the location to authorized emergency partners for the purpose of helping emergency services locate you. Your location is sent directly from your phone to emergency partners, not through Google.

After a completed emergency call or text during which ELS was active, your phone sends usage and analytics data to Google for the purpose of analyzing how well ELS is working. This information does not identify you and Google does not use it to identify you.

Note: Sending your location with ELS is different from sharing it via Google Maps.

Engage or disable searchbox effects

Press and hold on the searchbox at the bottom of the screen and a preferences box will appear. Within this is the option to enable or disable special effects. No one knows what the special effects are, but this is where to find them.

Get calendar and travel details at top of your home screen

The At a Glance feature will let you get calendar entries and travel information from

Google onto your home screen so they are easy to see. Long press on your wallpaper on the home screen and tap "home screen settings". Here you'll find the option to turn on the information you want - calendar, flights, traffic.

Have your phone automatically recognize songs

Introduced on the Pixel 2, this is a local feature on the Pixel 3 (so works offline), letting the phone listen to songs playing nearby and put the details on your lock screen. Head into settings > sounds > Now Playing to turn it on. You can also enable notifications for Now Playing.

HDR+ Processing

Google has deployed a proprietary technology that reduces "noise" and enlivens colors. In reality, it's taking a burst of shots, then combining the best of each image to create one final version of that image. Oddly enough, it's not automatically enabled. You don't need to go into the camera's advanced settings to enable the HDR+ control, however, as it's on by default. If you want manual control, however, you can

Quick Launch Camera

There's no need to fumble about to launch the camera; Google created Quick Launch to enable one-step photography. You need only double press the power/standby button to quick launch the camera. Go to **Settings** > **System** > **Gestures**, then tap jump to camera to enable it, allowing quick access from any screen.

Free Photo Backup

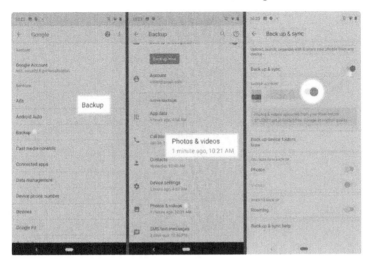

One benefit to buying a Google Pixel phone is Google provides you with free, unlimited, original quality backups of your photos and videos using the Google Photos app. To enable this free backup, open the Photos app and enable backup. It'll be the first prompt you see. Another way to access this feature is in **Settings** > **Google** > **Backup** > **Photos** & **videos**.

Other Photographic Upgrades

All of the following functions are automatically enabled. You should try using each one the first time you use the camera to gain an appreciation of all of the photographic tools the Pixel has:

- **Photobooth**: Waits for you to smile or make a funny face before snapping a selfie.

- **Topshot**: Takes multiple frames a few seconds before and after you take the shot, then suggests the best one.

- **Playground**: Opens an AR sticker feature that inserts characters like Marvel's Iron Man in pictures of you or your subjects.

- **Super Res Zoom**: Allows the user to zoom in on a subject without compromising quality.

- **Motion Auto Focus**: Tap a moving subject and the Pixel will automatically track it and keep it in focus. Great for shots of pets and children.

Call Screening

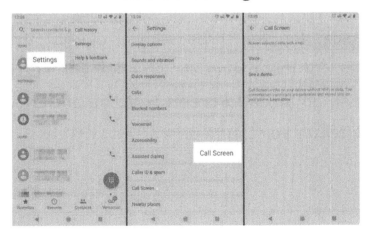

One of the cooler AI features is a very powerful real-time voicemail feature. When you receive a call, you're given an option to tap Call Screen. The caller will be asked to say why they're calling and the answer will be transcribed on your phone in real time. Then, you can choose to take the call, have Google Assistant read one of several pre-set statements, or immediately mark the number as spam.

Because this option is set to default, there's nothing you need to do to enable it, but you can revise it. For example, you can change the voice; tap the three vertical dots at the top right of the phone app to enter the **settings**, then go to **Call Screen** > **Voice** and choose a different voice option.

Night Light

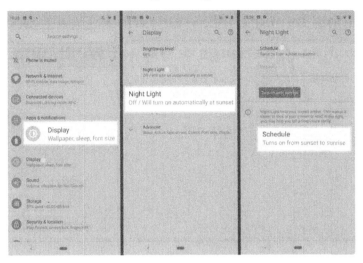

Blue light is not your friend when you're trying to sleep. Most phone manufacturers now offer a blue light filter that can be scheduled to kick in near bedtime.

Go to **Settings** > **Display** > **Night** Light, then tap Schedule. From there, you can choose to set a time for blue light to turn on or automatically align it with sunset and sunrise in your area.

You may also want to consider including Night Light as part of your Wind Down bedtime routine in **Settings** > **Digital Wellbeing.**

Dark Mode

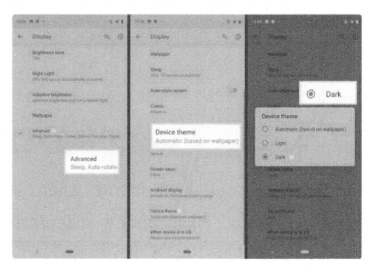

Android Pie 9.0 introduces a dark mode for the display to help with battery efficiency, and makes the Pixel more readable. The feature can be enabled 24/7, or you can set it so your wallpaper determines the color of the shortcut and notification shade. Go to **Settings** > **Display** > **Advanced** > **Device theme**, then tap **Dark**.

View your Now Playing history and put a shortcut on your home screen

Identifying songs is fine, but when you get home, you'll have forgotten what it was. Don't worry, your Pixel has you covered. Head into sounds > Now Playing > Now Playing History. This lists all the songs your phone heard and the time it heard them. You can click on a song to play it - Spotify, Play Music, YouTube, etc. You can also place a shortcut on the home screen to make it easy to get to this area. It's ace.

Enable or disable home screen rotation

Head into home settings > allow home screen rotation and you can view the home screen in landscape, rather than always viewing it in portrait.

Access Discover and customize it

Android has been pushing a page to the left of the home screen for many years. It was once

Google Now, now it's called Discover, a digest of topics you'll find interesting. Just swipe right to access it. In the top right-hand corner you'll find a settings menu where you can customize the content.

Turn off Discover/Google app

If you don't want this digest (above), access home setting > display Google app and you can turn it off.

Get the Pie dark theme

Head into settings > display > device theme. There are three options - light, dark and automatic. In the automatic mode, if you pick a black wallpaper the quick settings shade and apps tray also turn dark. It suits the Pixel 3 display really well.

How to help prevent your screen from cracking

- Use a protective case.

- Add a screen protector.

- Set your device on a flat surface when not in use so that it won't fall.

- Be cautious when letting young children use your device.

- Don't put your device near your keys or other hard objects that may scratch, crack, or crush the screen.

- Don't put your device in your back pocket where you might accidentally sit on it and crack the screen.

- Don't leave your device on vibrate near the edge of a surface, because it might fall when it vibrates.

- Don't expose your device to extreme temperatures, like leaving your device in your car on a very hot or cold day

HOW TO prevent water damage to your Pixel phone

Pixel 5 phones are designed to be water-resistant. But normal wear and tear can reduce water-resistance. To help your Pixel phone last longer, avoid actions that could lead to water damage.

Tip: When you're near water or steam, use a Bluetooth speaker to listen to your phone from a distance.

Keep your phone intact

Avoid chips or cracks in your phone's body or screen. Be cautious when letting young children use your phone.

Keep your phone dry

- Don't submerge your phone.

- Don't take your phone into a sink, shower, sauna, or bathtub.

- Don't take your phone into a swimming pool or body of water.

If your phone falls in water

1. Lift it out of the water.

2. Turn it off.

3. Dry it with a soft cloth.

4. Set it on a flat surface to finish drying at room temperature.

Clean your Pixel phone's back & sides

Wipe your phone clean when needed. Over time, a phone's back and sides could show stains or wear.

Wipe your phone clean

1. Turn off and unplug your phone.

2. With a soft cloth, wipe off your phone.

   o For most smudges or dust, use a dry cloth.

- o For most color transfers, like from makeup or a new pair of jeans, use a damp cloth.

- o When needed, use ordinary household soap or cleaning wipes.

How to make Your Phone Look its Best Longer

- Avoid dropping your phone.

- Avoid harsh cleaners and rough scrubbing.

- Avoid getting moisture or soap in openings.

- Avoid spraying cleaners or compressed air at your phone.

Tip: Use a case designed for your phone

How Google Play protect against harmful apps

Google Play Protect helps you keep your device safe and secure.

- It runs a safety check on apps from the Google Play Store before you download them.

- It checks your device for potentially harmful apps from other sources. These harmful apps are sometimes called malware.

- It warns you about any detected potentially harmful apps found, and removes known harmful apps from your device.

- It warns you about detected apps that violate or hiding or misrepresenting important information.

- It sends you privacy alerts about apps that can get user permissions to access your personal information.

How Google Play Protect works

Google Play Protect checks apps when you install them. It also periodically scans your device. If it finds a potentially harmful app, it might:

- Send you a notification. To remove the app, tap the notification, then tap **Uninstall**.

- Disable the app until you uninstall it.

- Remove the app automatically. In most cases, if a harmful app has been detected, you will get a notification saying the app was removed.

How malware protection works

To protect you against malicious third party software, URLs, and other security issues, Google may receive information about:

- Your device's network connections

- Potentially harmful URLs

- Operating system, and apps installed on your device through Google Play or other sources.

You may get a warning from Google about an app or URL that may be unsafe. The app or URL may be removed or blocked from installation by Google if it is known to be harmful to devices, data, or users.

You can choose to disable some of these protections in your device settings. But Google may continue to receive information about apps installed through Google Play, and apps installed on your device from other sources may continue to be checked for security issues without sending information to Google.

How Privacy alerts work

Google Play Protect will alert you if an app is removed from the Google Play Store because the app may access your personal information and you'll have an option to uninstall the app.

Send unknown apps to Google

If you choose to install apps from unknown sources outside of the Google Play Store, turning on the "Improve harmful app detection" setting will allow Google Play Protect to send unknown apps to Google to protect you from harmful apps.

1. On your Android phone or tablet, open the Google Play Store app   .

Tap Menu ≡ › **Play Protect** › Settings⚙.

2. Turn **Improve harmful app detection** on or off.

Uninstall problem apps

If an app still uses too much battery after you've force stopped and restarted it, you can uninstall it.

1. Touch and hold the app that you want to uninstall.

2. To see your options, start dragging the app.

3. Drag the app to **Uninstall** at the top of the screen. If you don't see "Uninstall," you can't uninstall the app.

4. Lift your finger.

Tip: You can see **Uninstall**, **Remove**, or both. "Uninstall" removes the app from your phone. "Remove" removes it from your Home screen only.

Restart your phone ("reboot")

To restart your phone:

- Press your phone's power button. On the screen, tap Restart↺.

- If you don't see "Restart," press your phone's power button for about 30 seconds, until your phone restarts.

Check for app updates

App updates can bring improvements that could fix your issue.

To see and get updates for your apps:

1. On your phone, open the Google Play Store app ▷ .

2. Tap Menu ≡ ❯ **My apps & games**.

3. Apps with available updates are labeled "Update."

   ○ If an update is available, tap **Update**.

   ○ If more updates are available, tap **Update all**.

4. Reset to factory settings

5. To remove any processes on your phone that could be causing the issue, you can reset your phone to factory settings.

6. **IMPORTANT: A factory data reset will remove all data from your phone.** While any data stored in your Google Account will be restored, all apps and their associated data will be uninstalled. Before you perform a factory data reset, we recommend backing up your phone.

7. If an app that you downloaded caused the issue and you reinstall that app, the problem could come back.

Know the Google Account username & password on your phone

To restore your data after resetting, you'll need to enter security information. Entering the information shows that you or someone you trust did the reset.

Make sure you have the security information before you reset your phone.

1. Be sure that you know a Google Account on the phone.

    1. Open your phone's Settings app.

    2. Tap **Accounts**.

    3. See a Google Account username.

2. Be sure that you know the password for the Google Account on the phone. To confirm, sign in to that account on another device or computer. If you don't remember the password, get sign-in help.

3. If you've set a screen lock, be sure that you know your phone's PIN, pattern, or password.

Tip: If you recently reset your Google Account password, wait 24 hours before performing a factory reset.

Back up your data to your Google Account

A factory data reset will erase your data from the phone. While data stored in your Google Account can be restored, all apps and their data will be uninstalled.

To be ready to restore your data, make sure that it's in your Google Account.

Plug in & connect

A factory reset can take up to an hour.

1. Plug your phone into a power source. Keep it plugged in until the reset completes.

2. Connect your phone to Wi-Fi or your mobile network. When the reset completes, you'll need to be connected to sign in to your Google Account.

Factory reset your phone

1. Open your phone's Settings app.

2. Tap **System** ⟩ **Advanced** ⟩ **Reset options**.

3. Tap **Erase all data (factory reset)** ⟩ **Reset phone**. If needed, enter your PIN, pattern, or password.

4. To erase all data from your phone's internal storage, tap **Erase everything**.

5. When your phone has finished erasing, pick the option to restart.

6. Set up your phone and restore your backed-up data

What sync does?

When your phone syncs, your Google apps refresh their data, and you get notifications about updates.

See your Google apps that can auto-sync

1. Open your phone's Settings app.

2. Tap **Accounts**.

3. If you have more than one account on your device, tap the one you want.

4. Tap **Account sync**.

5. See a list of your Google apps and when they last synced.

6. Check your other apps that couldn't sync

7. If an app doesn't show in your phone's Settings app under "Accounts," it can't auto-sync with your Google Account.

8. For these other apps, look in each app's settings menu for an option to sign in or sync.

Turn off auto-sync

You can turn off automatic syncing for certain apps made by Google, or for your whole Google Account.

Turning off auto-sync can help save battery life. To start auto-sync again after your battery recharges, turn it back on.

Turn off auto-sync for certain Google apps

1. Open your phone's Settings app.

2. Tap **Accounts**.

3. If you have more than one account on your phone, tap the one you want.

4. Tap **Account sync**.

5. Turn off the apps you don't want to auto-sync.

Note: Turning off auto-sync for an app doesn't remove the app. It only stops the app from automatically refreshing your data.

Turn off auto-sync for your Google Account

1. Open your phone's Settings app.

2. Tap **Accounts**.

3. Turn off **Automatically sync data**.

Manually sync your account

Manual sync refreshes your account data for all your apps made by Google, including any with auto-sync turned off.

1. Open your phone's Settings app.

2. Tap **Accounts**.

3. If you have more than one account on your phone, tap the one you want.

4. Tap **Account sync**.

5. Tap More ⋮ › **Sync now**.

Take care of your battery

Use the power adapter that came with your phone Other power adapters and chargers can charge slowly, not at all, or damage your phone or battery.

- Keep it cool

Avoid situations where your phone can overheat. Your battery will drain much faster when it's hot, even if you're not using it. This kind of drain can damage your battery.

Tip: Your phone warms up when it's plugged in, so don't keep it charging more than needed.

- **Charge as much or little as needed**

You don't need to teach your phone how much capacity the battery has by going from full to zero charge, or from zero to full charge.

- **Choose settings that use less battery**
- **Change screen display**

Let your screen turn off sooner. To reduce battery drain when you're not using your screen, set a shorter time before your screen turns off.

Open your phone's Settings app.

Tap Display › Advanced › Screen timeout.

Pick a time. We recommend 30 seconds.

- **Reduce screen brightness**

To save battery, lower your screen's brightness.

1. Swipe down with two fingers from the top of your screen.

2. At the top of the screen, move the slider left.

- **Have screen brightness change automatically**

To save battery, have your screen's brightness change with the light around you.

1. Open your phone's Settings app.

2. Tap **Display** › **Adaptive brightness**.

3. Turn on **Adaptive brightness**.

- **Turn off notification light**

If your device has an LED notification light, you can save battery life by turning off that light.

- **Turn off keyboard sound & vibration**

You can save battery life by turning off your device's keyboard sound and vibration.

- **Reduce background battery drain**

Restrict apps that use more battery

You can restrict background battery use by individual apps. In your battery settings:

- Your phone could recommend apps to restrict.

- You can see which apps use the most battery.

- **Keep adaptive battery & battery optimization on**

To help apps use your phone's battery only when you need them to:

- Keep adaptive battery on for your phone.

- Keep battery optimization on for all apps.

- Turn off high-drain features

- You can save battery by turning off tethering and hotspots when you're not using them You can let Wi-Fi turn off when your screen is off

- **Delete unused accounts**

Having fewer accounts on a phone can save battery. The phone's owner can delete accounts and user profiles.

- **Clean your Pixel Case fabric**

Hand-wash your Pixel Case fabric when needed.

Hand-wash your case

1. Remove your case from your phone.

2. Dampen a cloth or sponge with water.

3. Add a little mild soap.

4. With the damp cloth, rub your case's fabric gently in circles.

5. Let your case air dry at room temperature.

Tip: If needed, first use a stain remover pen. Avoid harsh cleaners, rough scrubbing, or soaking for a long time.

Customize Face Unlock

Now you can adjust how the face unlock feature works. Head to Settings -> Security -> Face unlock and enter your PIN when prompted. Here, you have a few options.

First, you can decide whether or not facial recognition unlocks your phone. If you want, you can disable the first toggle and the face scanner will only be used to sign into apps and authenticate you for payments. For those looking to be the most secure, this is the way to go, because, unlike Apple's Face ID, you can't close your eyes to prevent access. So if someone forces the phones toward your face, they are in.

The second option is "App sign-in & payments." This toggle manages whether you can authenticate with face unlock in apps and Google Pay and Play Store payments. Just as with unlocking your phone, it's a matter of security vs. convenience. For the strongest level of protection, disable this toggle. If it is disabled, to access secure apps, the person needs to know your password. This would also apply to making purchases on your phone. But, for convenience and solid security, leave it enabled.

A related option is the "Always require confirmation" toggle (don't worry, we'll go back to "Skip lock screen" in a second). This is turned off by default. When enabled, any app that uses facial authentication will require you to tap a button to confirm you wanted to authenticate. This is an extra layer of protection against accidentally unlocking an app or making a payment. Do note that some apps (such as Signal Private Messenger) require confirmation even if this toggle is off.

Finally, there is "Skip lock screen," a feature that many Face ID iPhone users having been begging for. Instead of having to swipe up after your Pixel 5 authenticates you with face unlock, you'll be taken right to the last app you were using. This could pose privacy risks, but it makes unlocking your phone nearly as quick as less secure face unlock methods (such as using the front-facing camera).

Google Pixel Camera Features

The camera that comes with Google Pixel and Pixel XL doesn't seem like much but it is really quite powerful. The reason why it doesn't seem so great is because most of the features are tucked away where you can find them so easily. Once you learn the tricks to the camera, you will quickly see how powerful it is for taking high quality images and even videos.

- Quickly open the camera by double pressing on the power button. When you need to take a shot quickly before the opportunity disappears, this feature will get that job done for you. This will open the camera even when the phone is in sleep mode.

- Twist your wrist twice to take a selfie. You don't need to press the selfie button and then pull the camera away from your face to take a good picture. Just twist your wrist to go to selfie mode while your hand is already pulled away from you for that perfect shot. Twist again to go back to regular camera mode.

- Take burst photos by holding down the "take photo" button. The camera will continue to take pictures for as long as you hold the button. When you let go you will see a gif of all pictures and that gif will save if you want to keep it. You can also see all the pictures you took together so you can keep the best one and delete the rest.

• The camera has a lens blur feature that lets you take beautiful pictures of up close objects. The feature focuses on the object in front and blurs out the background.

• Press down on the screen when you are about to take a photo to adjust the focus.

• Press down on the screen to see the slider to adjust brightness and white balance.

Enable app notification dots

This was a new feature in Oreo that lets you have a dot on apps that have a notification or something to show you. It's in Android Pie too. Head into Settings > Apps & notifications > Notifications and you'll see the toggle to turn on notification dots. Or you can long press on the wallpaper and hit "home settings".

Digital Wellbeing

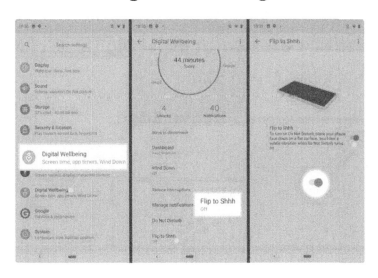

Google is leading the way in helping people control their smartphone usage by offering an array of digital well-being settings. This function displays the daily or hourly view of the time spent on your phone, how frequently you use different apps, and how many notifications you receive. It also offers tools enabling you to reduce interruptions, set a bedtime schedule, activate blue light, or turn the display to a gray tone.

Flip to Shhh is one such tool to check out. As the name suggests, when you place your phone face down it activates a Do Not Disturb status. Go to **Settings** > **Digital Wellbeing** to enable Flip to Shhh.

Active Edge

This is another one-step process that comes in quite handy. Want to turn on Google Assistant without touching the screen? Use Active Edge. Go to **System** > **Gestures**, then tap **Active Edge** to enable it. Once activated, it opens Google Assistant whenever you squeeze the long sides of the phone, even when the phone is asleep. In the settings, you can choose how hard you need to squeeze the phone, and you can use this same function to silence alarms and timers.

Google Sounds

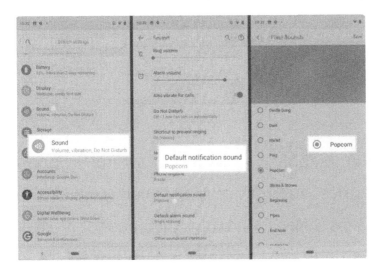

Pixel Sounds is the audio equivalent of a display app. It offers an enhanced variety of sounds in a bunch of different categories. Access the Sounds app by going to **Settings** > **Sound** > **Advanced** > **Default Notification**. This allows you to change your ringtone or notification sounds.

Home Screen Rotation

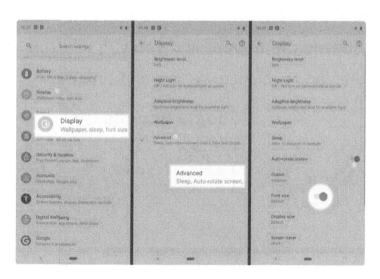

Remarkably, this useful function comes automatically disabled. One of the first steps after unboxing this phone is to go to Settings > **Display** > **Advanced** > **Auto-**

rotate Screen, then toggle the setting to ON. Once enabled, the home screen will display in landscape when held horizontally.

Always On

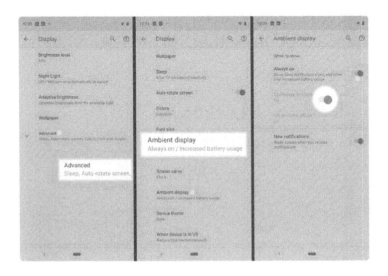

If you enjoy having the screen always show the time, date, and weather on your lock screen, you can make that happen with Always On. Head to **Settings** > **Display** > **Advanced** > **Ambient Display**. Here you'll find the option for the always-on display.

One nice touch is you can also include Now Playing, Google's song ID feature, on the lock screen. You can find this feature under Notifications. You might also consider enabling Double-tap to check phone and Lift to check phone, although either of these options is probably sufficient.

Lockdown

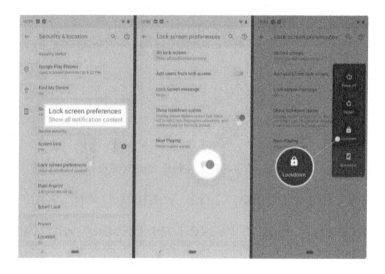

Security is strongly enhanced in the new Pixel OS; it now includes convenient secure-unlock options like the fingerprint sensor and Smart Lock. You can go one step further with

Lockdown mode, which comes in handy when you're traveling or in an unfamiliar place, as it deactivates certain features to make your phone more secure.

To enable Lockdown, go to **Settings** > **Security** and **Location** > **Lock screen preferences** > **Show Lockdown Option**, then tap the toggle to enable it; hold the power button and you'll see Lockdown as an option alongside shut down and restart.

Lockdown also allows you to temporarily disable the Fingerprint Unlock, Smart Lock, and Notifications on the lock screen. So now, in order to unlock the phone, you'll need to use either the PIN or a pre-programmed pattern.

Use app shortcuts

With Android Pie certain apps have shortcuts to actions that you can access by pressing and holding their icon on the home screen. This can be taking a video or photo with a camera, navigating home with Maps, or adding contacts, plus many more. Just press and hold and it will pop up.

Create shortcut icons

Once you have your list of app shortcuts pop up on the screen as above, you can drag and place them on the screen as their own individual icons. For example, on the camera, you can drag out a shortcut to go straight to the selfie camera.

Tips for Nighttime Photography

Taking dramatic nighttime photographs with your DSLR camera requires a different mindset and special techniques compared to daytime shooting.

- **Turn off the Flash for Night Photography**

If you leave your camera in Auto mode, it will try to fire the pop-up flash to compensate for the low light. All this will achieve is an "over-lit" foreground, with a background that's been plunged into darkness. Using any of the other camera modes will negate this problem.

- **Use a Tripod**

You will need to use long exposures to get great nighttime shots and that means that you will need a tripod.

If your tripod is a bit flimsy, hang a heavy bag from the center section to keep it from blowing around in the wind. Even the slightest amount of wind can shake the tripod while exposing and you may not be able to see a soft blur on the LCD screen. Err on the side of caution.

- **Use the Self-Timer**

Just pressing the shutter button can cause camera shake, even with a tripod. Use your camera's self-timer function, in conjunction with the mirror lock-up function (if you have this on your DSLR), to prevent blurry photos.

A shutter release or remote trigger is another option and a good investment for any photographer who takes long exposures on a regular basis. Purchase one that is dedicated to your model of camera.

- **Use a Long Exposure**

To create great nighttime shots, you need to allow the dim ambient light to sufficiently reach the image sensor — so you'll be shooting longer exposures.

A minimum of 30 seconds is a good place to begin and the exposure can be extended from there if necessary. At 30 seconds, any moving lit objects in your shot, such as cars, will be transformed into stylish trails of light.

If the exposure is very long, then it may be out of your camera's range of shutter speeds. Many DSLRs can go as long as 30 seconds, but that may be it. If you need a longer exposure, use the bulb setting to keep the shutter open as long as the shutter button is pressed. A shutter release is essential for this step; they typically include a lock so you do not have to actually hold the button the entire time (just don't lose it in the dark!).

The camera will take longer to render and process these long exposures. Be patient and let it process one image before trying to take the next one. Night photography is a slow process and, besides, you want to see the capture on the LCD screen so you can adjust the next exposure to perfect the shot.

Tip: *The camera will take longer to render and process these long exposures. Be patient and let it process one image before trying to take the next one. Night photography is a slow process and, besides, you want to see the capture on the LCD screen so you can adjust the next exposure to perfect the shot.*

- **Switch to Manual Focus**

Even the best cameras and lenses have a difficult time with autofocus.If you even have a hard time finding something to focus on in the dark, use the distance scale on the lens. Estimate how far away a subject is in feet or meters, then use a flashlight to see and set that measurement on the lens.

If the only subject is very far away, set the lens to infinity and stop down as far as the lens will go (a minimum of f/16) and everything should fall into focus. You can always check on your LCD screen and adjust the next shot accordingly.

- **Increase the Depth of Field**

A large depth of field is best for nighttime shots, particularly when you're photographing buildings and lit structures. A minimum of f/11 should be used though f/16 and up are even better — although less light enters the lens, so adjust your shutter speed accordingly.

For every f/stop move you make, your exposure will double. If you shot at f/11 for 30 seconds, then you will need to expose for a full minute when shooting at f/16. If you want to go to f/22, then your exposure would be 2 minutes. Use the timer on your phone if your camera does not reach these times.

- **Watch Your ISO**

If you have adjusted your shutter speed and aperture, and still do not have enough light in your photograph, you could consider upping your ISO setting to help you to shoot in lower light conditions.

Remember, though, that a higher ISO will also add noise to your image. Noise makes its biggest appearance in the shadows and night photography is filled with shadows. Use the lowest ISO you can get away with!

- **Have Spare Batteries on Hand**

Long exposures drain camera batteries. Carry spare batteries if you plan to conduct a lot of nighttime shots.

- **Experiment   With Shutter   and Aperture Priority Modes**

If you want to help yourself learn as you go along, consider experimenting with these two modes. AV (or A — aperture priority mode) allows you to choose the aperture, and TV (or S — shutter priority mode) lets you choose the shutter speed. The camera will sort out the rest.

Priority-mode shooting is a great way to learn how the camera exposes images, and it will help you to achieve the correct exposure.

Manage quick settings icons

In Android 9 you can manage the order of the quick settings tiles by dropping down the usual shade from the top of the screen and hitting the pencil icon at the bottom to edit. Now you can re-order, add or remove new quick access toggles, making it easier to get the controls you want.

Quickly select a Wi-Fi network

Swipe down for Quick Settings, then press and hold the Wi-Fi icon. This will go directly to the Wi-Fi settings, it's great when you can't figure out what's going on with Wi-Fi.

Quickly manage Bluetooth

The same applies to Bluetooth. Swipe down the Quick Settings shade and press and hold the Bluetooth icon. If you're failing to connect to your car, you can instantly see what's going on.

Turn on torch/flashlight

There's no need for a separate app, just tap the button in Quick Settings to turn on your flash as a torch. Or just say "Ok Google, turn on torch/flashlight" and it will turn on.

Cast your screen

Want your Android device on your TV? Just swipe down and tap Cast screen and it will be sent to your Chromecast. If it's not there, add the Cast tile to your Quick Settings using the method mentioned above. Not all apps are supported though.

Display tips and tricks

Turn on always-on display: Head into Settings > Display > Advanced > Ambient display. Here you'll find the option for the always-on display, which will show the time, date, weather on your lock screen. You can turn it off to save battery life.

Turn on double tap to wake

This has been on a number of devices previously, but is now a standard Android feature. Head into Settings > Display > Advanced > Ambient display and tap on "double-tap to check phone". This only works when the always-on display (above) is turned off. Then you just double tap the screen and you'll be shown the details.

Get notifications when you lift your phone

Head into Settings > Display > Advanced >Ambient display and you can turn on the option to show you the always-on display when you lift your phone up. That means you can glance at the time and your notification icons, without having to press any buttons or anything.

Wake the display when new notifications arrive

If want the display to fully wake up when you get a new notification, this option is also in the ambient display settings (as above). You'll need to make sure you're not getting overwhelmed with notifications, or it will drain your battery a little faster.

Manage the colors of the display

This has become a big deal since the controversy surrounding the Pixel 2 XL color hue. Head into settings > display > colors and you'll find the options offered - natural, boosted or adaptive. We've found adaptive to be the best for most use cases.

Have night light automatically turn on/off at dusk and dawn

Night light aims to reduce the blue light from the display to make it better for viewing at night, reducing the brightness and the strain on your eyes. Head into Settings > Display > Night Light and you'll find all the controls. in the schedule you can customize when this happens, with automatic sunset to sunrise being an option.

Change the hue of Night Light

If you want to change the color tone of Night Light, head into the settings as above and you can change the intensity. If you find yourself regularly turning it off because it's too yellow, you could probably make it better with a hue tweak here.

Camera and photos tips

Pixel basically means camera these days and the Pixel 3 camera doesn't disappoint. It's getting a little more complicated, so here are the tips you need to get it singing sweetly.

Quick launch the camera

Double press the power/standby button to quick launch the camera, it's a great feature. The settings for this control live in Settings > System > Gestures. Here you can turn on "jump to camera" to allow quick access from any screen.

Swipe between photos, video, other camera modes

You can swipe from photo to video capture and to other modes in the camera viewfinder, which you might prefer to hit the buttons. Simply swipe up or down the screen in landscape, or left and right in portrait and you'll switch from photo to video capture.

Find the camera settings

These are no longer visible from the main camera view. As above, swipe across to "more" and tap on that option. There you'll find the settings.

Instant zoom

If you want to instantly zoom in on something and you've only got one hand free, just double tap anywhere in the viewfinder and the camera will jump to 2x zoom. This is great if you don't have a free hand to use the slider, but it's not full zoom - you can then zoom in further if you wish.

Turn off the shutter sound

That noise is pretty annoying, right? As we mentioned above, swipe across to More > Settings and you'll see the toggle for camera sounds.

Use Night Sight

One of the most impressive features of Google's Pixel Camera is a feature called Night Sight. It effectively lets you take photos at night without a flash and still produce an image that's bright, colorful and crystal-clear.

It does this using some really powerful software normally found on only the most expensive phones.

To turn on Night Sight, simply open the camera and swipe along to More. Tap on Night Sight. Now line up your shot and press the shutter button. An important thing to note is that you'll need to hold the camera extremely still for a second or so after you've taken the picture, so while not essential, a tripod will really help.

Night Sight actually works really well during the daytime too - the key here is to experiment and see what works.

Use Smartburst and Top Shot to capture great moving action

Press and hold the shutter button and the Pixel 3 will rattle off lots of photos. Firstly, you can manually select the one with the picture you want or you can open the Google Photos to view your photos and tap the burst button bottom centre. This will have the option to only show the best photos from the burst, using Google's AI to give you the best.

Take burst photos with automatic animation

Google Photos has a great auto-animate feature which uses bursts of photos and turns them into animation. It's great for capturing not only a photo of some action, but all the activity that surrounded it. Capture the action as described above and Google Photos will automatically turn it into an animation once it recognizes a series of photos. If it doesn't do it automatically, you can force the animation to be created via the burst button in Google Photos as above.

Adjust the exposure compensation

Exposure compensation lets you lighten or darken a scene when the automatic metering doesn't quite get it right. For example, an illuminated subject on stage in a dark theatre will often automatically over-expose. Dial down the exposure and the dark part of the room will darken, returning to a more dynamic picture. Simply tap on what you want to focus on (your subject) and then on you'll see the brightness scale appear on screen. Simply drag this up or down accordingly to get the result you want.

Lock the exposure and the focus

This is a trick used by photographers to make sure that the camera locks onto the correct exposure and focus for a subject in the frame and keeps that until the photo is taken. It's useful, for example when there's a lot going on that the camera might focus on instead, perhaps things moving elsewhere in the frame. On the Pixel 3 when you tap to focus there's lock icon at the top of the exposure slider - tap this to lock.

Enable/disable Motion Photos

Like Apple's Live Photos, when you snap a photo you can have it capture a short burst of video. To enable or disable it, tap the small icon that looks like a solid circle inside a ring. You'll also find this icon in Photos app on any images that were snapped using the Motion Photo feature.

Add a manual HDR+ switch

Google Pixel 3 and Pixel 3 XL take really great photos thanks to Google's automatic HDR+ technology. If you'd rather it wasn't automatic, you can add a button to switch it on or off by heading to the camera app, open the more menu, hit settings > advanced and toggle the switch.

Get Google Lens suggestions

This is a really clever option that will highlight certain information via the camera. Just point the camera at a phone number, name or website and a link will be offered to open Chrome, place a call or open up your Contacts with that person. It's on by default, but you can find it in "more" > settings > Google Lens suggestions.

Engage Google Lens through the camera

Google Lens is an AI system that identifies objects and gives you information. You can find it in the "more" option on the camera, or you can get to it by pressing and holding in the viewfinder. Then then flips to Lens and find things for you.

Engage portrait mode

Craving that blurred background effect? Just swipe to Portrait. Then you simply have to line up your subject and take the picture. It works on both the front and back cameras.

Zoom out on the front camera for a wefie

The Pixel 3 has a wide-angle camera on the front. Simply flip to the front camera and pinch and you'll switch to the wider camera.

Engage beauty mode

Ok, it's not called beauty mode, it's called "face retouching". Hit the icon on the side with a little face and you get the option of natural and soft - or to turn it off. This can also be use in conjunction with portrait mode for the ultimate selfie. Face retouching can also be used when in portrait mode on the rear camera to make other people look better.

Engage video stabilization

Head into the settings menu and you'll find the option to turn on video stabilization.

Explore the More menu a little more

Anything that's not a main camera mode is in the more menu at the end. Here you'll find Photobooth which is fun for selfies, Photo Sphere which is a hangover from Nexus days and lets you capture 360 photos, as well as Playground which will drop AR characters into a photo. It's all a lot of fun and worth exploring.

Lock the video camera to 30fps

The Pixel camera has an auto FPS mode (when in the default 1080p) that will switch up to 60fps if it sees a reason to - for fast moving action. It can make this switch during a video, changing the frame rate. The aim is probably to give smoother results on playback via the device, but you do get the option to lock to 1080/30p - which might be useful for video makers. You'll see the icon bottom left in the viewfinder in video mode.

Split-screen multitasking

Android offers split-screen multitasking and it now uses Overview to control it. Swipe up to pop into Overview, then tap the app icon at the top and you'll find "split screen as an option". Tap this and it will move to the top of the screen. You can then scroll through Overview to find the second app, or open another app and it will take up the bottom of the screen.

To return to single screen/not split

If you find yourself stuck in split-screen, press the home button. If there's still an app at the top, swipe it down and it will return to full screen. Then press the home button again and you're back to normal.

Change the default app

Android lets you decide which is the default app, if you have more than one that will do the same thing. Under Settings > Apps & Notifications > Advanced you'll see the default apps option. Here you can set your default browser, launcher, SMS app and so on.

Control app permissions

Android lets you manage all the permissions for each app on an individual basis. Go to Apps & notifications, select the app and hit Permissions. This will let you toggle permissions on and off, so you can disable location access, for example.

Access Google Play Protect

This is Google's app scanning feature. If you want to find it open the Google Play app and you'll see it at the top of the "my apps & games list".

Disable picture-in-picture

Picture-in-picture will allow a thumbnail version of an app or video to play once you return to the home screen. That's great, but if you don't want it, head into apps & notifications > advanced > special app access > picture-in-picture. Here you can toggle off apps you don't want using it.

Worried about you app usage? Digital Wellbeing will help you

If you're worried about how much time you spend on your phone, then head into settings and find Digital Wellbeing. This will not only give you a breakdown of your app and phone usage, but you can set timers to help you. There's also the option to add a shortcut to your app tray so it's easy to get to.

Volume tips and tricks

Notifications on Android are the best around, giving you loads of option and loads of control. But there are so many options it can get confusing.

Direct reply

With recent versions of Android you'll often be able to direct reply from any app that has it built in. Swipe down on any notification card and if there's a "reply" option, hit it and type away without leaving the screen. Sometimes the toast notifications will give you the direct reply option too, so you can reply when you're playing a game without taking your eye off the action.

Quickly switch to vibrate alerts

If you want silence, but are after vibration alerts still, then push the volume button and tap the bell on the pop-up at the side. This will switch to vibrate.

Turn down media volume

Hit the volume up or down button, and the volume slider will appear on the right-hand side. Tap the settings cog and you will access all the volume controls. Here you can turn down media volume.

Squeeze to silence alarms and calls

You can quickly silence your phone with a squeeze. Head into settings > system > gestures > active edge. At the bottom of this list you'll find the option to squeeze for silence.

Engage Do not Disturb

Swipe down Quick Settings and tap the Do Not Disturb icon. You'll be notification free.

Schedule Do not Disturb

Swipe down Quick Settings then press and hold the Do Not Disturb button. Choose Schedule > Turn on automatically and you'll find the automatic rules. Here you can set times for Do not Disturb to automatically turn on and off, like evenings or weekends.

To turn off notifications on an app

Go to Settings > Apps & notifications > Tap on the app you want. In Notifications you can block all notifications for any app on your device. Or, when you see a notification you don't want, slowly swipe it right to reveal a settings cog. Hit that and you'll be able to block notifications from that app.

Hide sensitive information in lock screen notifications

You can have lock screen notifications without too much information being revealed. Head to Settings > Security & location > Lock screen preferences. Here you can set the phone to hide information so it can't be read by everyone.

Have Now Playing recognised music appear in the notifications tray

You can opt to have recognized music appear in the notifications tray (or remove it if you don't like it). Head into settings > apps & notifications and locate Pixel ambient services. Within this app is the option to control the messages that you'll get then music is recognized.

Google Assistant tips and tricks

Google Assistant is getting into all parts of Google's devices, expanding its feature set and powers with machine learning and AI taking over the world. Here's some great things to try with Google Assistant, but hit the link below for load more tips.

117

Squeeze to launch Google Assistant

Head into settings > system > gestures and you can control Active Edge, set the squeeze sensitivity, or disable it if you don't like it. You can also opt to use it when the screen is off. Squeezing will start Google Assistant listening so you can just start talking.

Launch Google Assistant

If you want to launch normally, press and hold the home button. Google Assistant will pop-up and take you through to the interface where you can talk to Assistant. You will also be served results that you can tap to get more information, or to move through to over apps. If you want to type instead of talk, tap the keyboard in the right-hand corner.

Swipe up Google Assistant to see more personal information

Swipe up once you've launched Google Assistant and you'll find a load more information waiting for you. You can see what's coming up or check your commute, for example.

Turn on the Ok Google hot word

When you setup your phone, you'll be prompted to setup the Ok Google hot word. If you choose not to, you can set it up at other times easily. Just unlock your phone and say Ok Google and the setup page will open.

Open an app with Google Assistant

Simply say "Ok Google, open Netflix" and it will open Netflix or any other app. It's smart too, as for some apps, Assistant can navigation content within them - like watching a specific show on Netflix, or playing a specific artist on Spotify.

I'm feeling lucky

If you're looking for Google Assistant's Easter Egg, trying saying "I'm feeling lucky". This will take you to a trivia quiz that's loads of fun.

Pixel Stand setup tips

For the first time the third generation of Pixels has wireless charging, and its own charging stand which does more than just topping up the battery. Using some clever connectivity trickery, the stand turns your phone into a bedside assistant.

Get to Pixel Stand settings

You'd think, having the Pixel Stand, its settings would be somewhere obvious. Turns out, it isn't. Drop-down your quick settings/notification shade and long press the Bluetooth icon. This takes you to "connected devices", now tap on "previously connected devices". Here, you'll find Pixel Stand and all you need to do now is tap the settings cog next to it.

Show photo slideshow

When you first set up the Pixel Stand it'll take you through some options, asking if you want to use it as a photo frame. If you want to change the Photo Frame settings, tap on "photo frame" within the option in the Pixel Stand settings. Now you can enable or disable the feature by toggling the "Photo Frame on Ambient display" option, as well as choose a Google Photos album you want it to cycle through.

Sunrise alarm

Tapping on the Sunrise Alarm option lets you customize how the phone behaves in the 15 minutes leading up to your morning alarm, when docked on the Pixel Stand. You can toggle it on and off, as well as choose the times between which it should be active. When active, it gradually turns from red to orange to yellow in the lead up to your alarm sounding.

Screen off when dark

One setting you can enable quickly by toggling it on, is the ability for your screen to turn black when it's dark.

Do not disturb while docked

One other feature that's a quick toggle on is the setting that switches Do Not Disturb on as soon as the phone is docked in the stand.

Tricks for your android phone:

Your Android smartphone is capable of a wide variety of things, but chances are there are a lot of things you had no idea your powerful little pocket PC could do.

From quick and simple ways to change your wireless network to an easy way to cast your phone screen to another device, we've got an assortment of things you probably didn't know your Android phone could do.

These 10 cool abilities will have you showing off your abilities to your friends and having them ask you where you learned it all – you can link them here and share the knowledge, of course!

1. Cast your phone screen to your favorite TV

All you need to share what's on your phone's screen with your television is a Chromecast or a television that's set up to work with Android devices and you can mirror what's on your screen there. This comes in handy if you want to share pictures or videos on your phone to someone else, watch YouTube without having to use an external app, or even play your favorite mobile games on a larger screen.

Go to the Quick Settings menu and choose Cast. Your device will pop up the Chromecast that's set up, or the television of your choice (if it's a smart TV) and you can start casting. It's very simple, and quick to set up.

2. Lock people out of specific apps

We've all been there - someone's asked to use our phone, and we've scrambled to hide certain pictures and information we don't want others to see. There's actually a way with Android devices to hide this with a few button presses so anyone you lend your phone to for a few moments will only be locked to one area until you enter your phone's lock screen code. That way, they can't use parts of your phone unless you put the code in again.

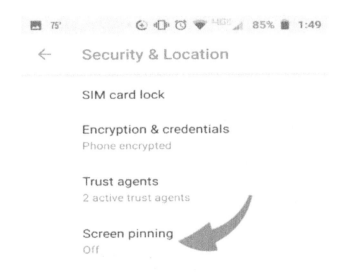

Go to Settings, then Security, and turn on
Screen Pinning. When turned on, open the app
that your friend or loved one needs to use.
Open Overview with the square button below
the phone screen. You'll see an icon that looks
like a pin in the lower right corner. Tap it, and
it'll remain pinned to the front. Voila! Now
that's all anyone can use until you enter the
password.

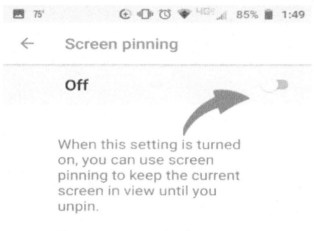

When this setting is turned
on, you can use screen
pinning to keep the current
screen in view until you
unpin.

To use screen pinning:

1. Make sure screen pinning
is turned on.

2. Open the screen you
want to pin.

3. Tap Overview.

4. Swipe up and then tap
the pin icon.

3. Enable one-handed mode

Some Android phone sizes can be particularly
unwieldy, depending on the model you use.
Luckily, you can opt for the special one-
handed mode on most Android phones using
one quick shortcut. It's on Google's keyboard,

which you'll need to get if you don't have it. You'll likely have it already as your default option if you use a Pixel or Nexus phone, but if you use a Samsung or LG phone, you'll need to download the keyboard to get things ready.

When you download the keyboard, open it up and tap and hold the backslash key. Pull up to the right-hand icon to turn on one-handed mode. There will be an arrow here that lets you position the keyboard from side to side and reposition it. The icon on the top will let you restore the keyboard to its full size. Now the keyboard can be used with one hand if you so desire.

4. Change out wireless networks quickly.

Sometimes you may need to swap between wireless networks in an expedient manner. Perhaps you're on the wrong network at a hotel or you want to use a friend's signal instead of your hotspot. Instead of going to Settings and then Wi-Fi, simply swipe down twice from the top of your screen and open the Quick Settings menu. Press on the name of the network you're connected to and you'll be given an entire list of networks all around you. You can now swap to an eligible network from there.

5. Swap into Priority Mode

You may never have heard of Android's Priority Mode, but it's an extremely useful feature. It lets you put your phone in "Do not disturb" mode while still letting certain notifications come through that you select. Essentially, you'll be able to choose who can bother you while everything else is snoozed.

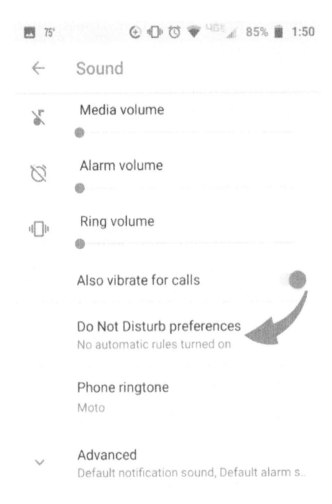

Start by going to Settings, then Sound & Notification, then Do not disturb. From there, change up the settings and notifications that you still want to let through. You can turn it off and on quickly via Quick Settings, but it will let through important folks, and that's what makes it so cool.

6. Read through notifications you accidentally closed

It can be easy to flippantly swipe away a notification that you meant to read. It's simple to bring one back, but your phone doesn't make it obvious. You can see all of your recent notifications if you'd like, actually. Just tap and hold on an empty part of your home screen. Select the Choose Widgets option, then look for Settings. Drag the icon to a space on your home screen and you'll see a list pop up. Select Notification Log, and then tap the icon. You'll now be able to scroll through all of your device's notifications.

7. Clear your default apps

Sometimes, you'll open a link that prompts the corresponding app to load, such as

YouTube, Facebook, or Twitter, when you meant to load it in the browser. You can clear the default app listing and keep this from happening in a very simple way so you can continue browsing without having to switch out apps when you didn't mean to in the first place.

Go to Settings, then Apps, and look for the app you want to stop opening up. Find Open By Default when you've located it, and choose Clear Defaults. No more pesky app loading!

8. Enable Smart Lock

Keeping your phone secure is always a priority, but when you're home and lounging around maybe you're not as interested as locking everything you can. You might want to try out Smart Lock, which takes away the requirement to use a keypad lock when in a trusted area like your house. You can opt to use your face, wireless networks, GPS locations, and more to prevent your phone from locking. This way you don't have to enter your password each time.

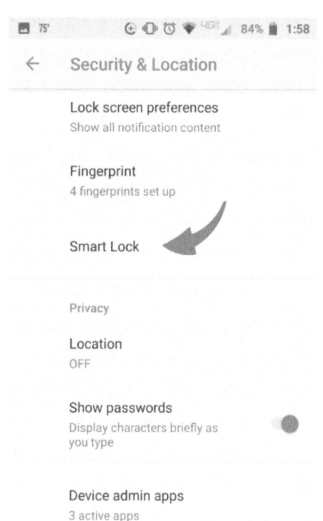

Go to Settings, then Security, then Smart Lock to get it set up, and bask in the light of simple phone usage when in trusted zones.

9. Change independent volume settings

When you go to change the sound on a certain aspect of your device, you're probably accidentally always turning one part down, but not the one you meant to, right? Maybe you turned down the phone's volume entirely, but you meant only to adjust your ringtone's volume.

Instead of relying on your volume buttons, tap the physical volume buttons on your phone, and look for the volume setting to show up on your screen. Look for an arrow on the box, tap it, and check out the volume sliders here to manually adjust the ones you want to change. Now you can do this quickly and easily.

10. Encrypt your phone's data

It's always a good idea to keep your data safe, even if you don't house much private information on your phone. You can actually manually enable encryption on your phone, which will make it unreadable by outside parties without a PIN or password to decrypt it. It's very simple to turn on: Go to Settings, then Security, then choose the Encrypt Phone option.

It can take a while to enable if your phone wasn't previously encrypted, and could potentially cause your phone to slow down if you're running an older version of Android, but it's an invaluable step to take if you want to protect the things already on your phone.

11. Put some smarts into your lock screen

You're good about keeping your Android phone locked down. You have a password, a

PIN, a pattern, or a fingerprint that's required to access your phone. That's a smart security move, but sometimes you want convenience. This is where the Android Smart Lock feature comes in. It lets you keep your phone or tablet unlocked in certain situations, like when you're near your home.

Head to your Settings and look for the Security menu. Choose Smart Lock and open it. This gives you a suite of options ranging from on-body detection to Voice Match, which lets you unlock the device with your voice.

If you're new to this, then you might want to start with the "Trusted places" setting and set up your home as the place to keep your phone unlocked. You could also set more locations, like your workplace, but keep in mind your phone will remain unlocked in those locations if you happen to leave it sitting somewhere.

Here's how to keep your family and friends from snooping on your phone.

12. Get your Quick Settings even faster

When you want to access your Wi-Fi, battery, do-not-disturb, Bluetooth, and other basic settings, you probably swipe down from the top of your phone. This shows the icons for some of your settings along with whatever active notifications you have. You have to then swipe down again to open the whole tray of settings.

But there's an even quicker way to get to your Quick Settings. Swiped down with two fingers from the top and, like magic, your Quick Settings are all laid out, ready for you to do whatever you need.

13. Swipe your way through your Chrome tabs

Most of us tend to have multiple tabs open in the Chrome browser whenever we're out exploring the internet. On a desktop or laptop, it's easy to click between tabs. On Android, however, you probably tap on the square at the top that shows the number of tabs you have open, and then sort through the pile of tabs until you find the one you want to switch to.

There's another way to move between your tabs in the Chrome app and it's more fun. Just swipe across the address bar either left or right to zip between your open tabs.

14. Use Split Screen

Starting with Android 7.0, users have been able to access a split-screen feature that lets two apps share the real estate on your screen simultaneously. Tap on the square icon at the bottom of your screen to see your running apps. You should see a message at the top that reads "Touch and hold a window, then drag here to use split screen."

For example, you can touch and hold on the Gmail window and wait a moment for it to pop out a little. Drag it to the top and it will take up half the space. Then, click on another running app, like Chrome, to fill in the bottom and you can now use both of them. Hold down the square icon (which should now look like two stacked rectangles) to return to a normal screen.

Keep in mind that not all apps support split-screen, so you may run into some that won't work with this feature. If you find split screen is too cramped for you, then read on for a different tip that will help you when you're using two apps at the same time.

15. Double tap the square

Here's one clever Android feature that's often overlooked. You're already very familiar with the three icons that are usually found at the bottom of your screen: the triangle, the circle, and the square. Touching the square shows all of your open apps stacked atop each other. That's great for choosing an app out of a whole line-up, but the square has another handy function. Double-tap it to automatically switch between apps. This is especially nice if you tend to keep a couple apps open and need to bounce back and forth between them.

Find your Android phone using Find My Device

The easiest way is to head into your Chrome browser and type "find my device". Google will return a window that will locate your Android devices using Find My Device. You'll have to log-in to access the details, but you'll then be told the location of your phone, the battery status and what Wi-Fi network it is connected to. You'll also have the option to erase, lock or play a sound. On the device you've located, it will have a notification to say it's been found.

Get pop-up/floating navigation

You can get Google Maps to give you a floating navigation map, so you can be browsing Twitter while you follow walking directions, saving you from constantly switching apps. Just start your navigation in Google Maps and hit the home button and Maps will shrink into a floating live window you can place where you want on the screen. You can control it with the picture-in-picture controls.

Check for Android updates

You want the latest version of the software, so head into settings > system > advanced > system updates. Here you can manually check for any updates that haven't been pushed. There probably won't be anything, but at least you know how to check.

Enable developer settings

To turn on the developer settings, head into settings > system > about phone. Scroll to the bottom and repeatedly tap on the Build number. After a number of taps, you'll unlock the developer options.

Turn off the developer options

There's no magic tapping for this. Once you've unlocked those options, a new section appears in the Settings menu. Open it up and there's a toggle switch at the top. Here you can turn it off, and that menu option vanishes.

Find the Android Pie easter egg

Pie's Easter Egg is a paint app. Head into settings > system > about phone. Then tap the Android version and a card pops up. Then tap Android 9 and you'll flip to the multicolored P. Then tap the P logo a few times and you'll launch into paint. You can then scrawl, change colors and pens and have a bit of fun.

Search settings

Rather than rooting through everything, you can search the settings. Just open up the Settings menu and there's a searchbar at the top. This can basically search any setting on the phone, so it's really easy.

Find the Google Settings

There was previously an app to handle Google-specific settings, in Pie this is in the main Settings menu. This is where you'll find settings for accounts and services, backup, and transferring content to a nearby device. It's an odd collection and there's a lot of duplication, so you'll find many of these settings in individual apps too.

Storage tips and tricks

The biggest difference between Pixel and other Android devices is that those non-Google phones will give you a microSD card slot, giving you a lot more flexibility.

Automatically clear backed-up photos

There's a Smart Storage option in Oreo that will automatically clear space on your phone by removing photo and video backups. For the Pixels you have free unlimited storage for these in Google Photos, so removing that duplication from your phone presents no problem. Head into Settings > Storage > Smart Storage. Here you can set the timeframe for removal - 30, 60 or 90 days, or you can do it right away.

Free up storage space

Android Pie makes this really easy. Head into Settings > Storage and you'll see a big button saying "free up space". That will then give you a list of things you could remove, like downloads you might no longer need, or apps you never use. The latter are arranged in size and dates so you can easily tick the box and hit delete.

See which apps are using up the most storage

If storage is getting to be a problem, head into Settings > Storage and you'll get a breakdown of categories for your storage. If you find something that looks much higher than you'd expect, it's worth checking out. For example, if you've downloaded a load of videos you've watched in the "movies & TV apps" you can remove them.

How to have Google Assistant screen your calls

If a call comes in that you suspect may be spam — or you just don't want to take it — tap

Screen call and Google Assistant will answer for you, saying, "Hi, the person you're calling is using a screening service from Google, and will get a copy of this conversation. Go ahead and say your name, and why you're calling."

You'll see what they say transcribed in real time on your Pixel screen and you can choose whether to answer, send a quick reply, or report as spam.

How to configure Ambient Display

We definitely recommend turning Ambient Display on, so that you can see new notifications and other information on your Pixel 3's screen even when it's locked, but you should configure it. Since battery life isn't the best, we wouldn't recommend leaving it on all the time. Go to Settings > Display > Advanced > Ambient Display and make sure that the Always on mode is off, but New notifications is on. We also turned on Double-tap to check phone and Lift to check phone, but you might feel one of those options is enough.

Never run out of space again with Smart Storage

The great advantage of choosing a Pixel phone is that everything on the device is securely backed up in the cloud using Google Drive.

Safe in the knowledge that your data is secure, Smart Storage can free up space on your phone by automatically clearing out photos and videos that are over 60 days old.

Don't panic: it's not deleting them. Instead it just removes them from the phone, leaving only the thumbnail behind. If you want to view them again, just tap on the thumbnail and your phone will instantly download the original back from the cloud.

To turn Smart Storage on, go to Settings and then tap on Storage and then turn Smart Storage on.

If you're still struggling for space remember that if you're a BT broadband customer you get free online storage through BT Cloud.

Recognize a song just by glancing at your phone

Ever heard a song playing in the background and wondered what it is?

Well rather than having to quickly scramble to open an app in time to catch it, Google's Pixel phones have a feature called Now Playing which intelligently knows what's playing and then displays it on the lock screen.

Even if you miss your chance, you can access a history of all the songs it picked up, so you can go back in time and trace the songs that were playing during a visit to a bar or on a TV show.

To turn Now Playing on go to Settings and then tap on Sound. Now tap on Now Playing and turn it on. If you want to be notified when it detects a new song just turn Notifications on.

Customize your home screen with widgets

Just like every other Android phone, the Pixel 3a's home screen is completely customizable, so you can keep it minimal or pack it full of information at a glance.

One of the ways that you can enhance your home screen is by adding Widgets - interactive elements that show you information (such as Spotify music controls, your week's calendar) without having to open that specific app.

Adding widgets is incredibly easy, simply go to the home screen and long press on any empty area. Now tap on Widgets and you'll be able to see all the widgets available.

Select the one you want by pressing and holding and then dragging it to where you want it on the home screen.

Some widgets can be resized. To change the size of a widget, press and hold on it once it has been placed and you'll see a white box with a white circle on each side. Press and hold on one of the circles and drag the box to resize it.

Use the Fingerprint Sensor to Quickly See Notifications

Google Pixel's Move features let you use your fingerprint to perform some actions with the phone. You have to turn this feature on in the Move settings but it will take only a few moments and you will end up with the most convenient feature the device has.

Once you have the fingerprint sensor locked into your fingerprint you can quickly turn on your phone, load up the home screen, and see your notifications with just a swipe. All you have to do is press the press the power button and then swipe down the sensor to start it all up without having to input a code.

Free Up Space Feature

Google Pixel and Pixel XL do not come with SD capability so you can't add space to the device. It does, however, provide free cloud storage with Google so that is better than an SD card. The problem is that you can quickly fill up your phone storage with those burst photos and other downloads.

The phone is set to automatically save all your files, photos, videos, and more to the cloud. It is a useful backup feature but there is another more hidden feature that makes this even better. The Free up Space feature will remove all duplicate photos or files from your phone that are already backed up on the cloud. You can do this manually when you run out of the space or set the phone to do this automatically. If you want to do it manually you can free up space on the entire device or just press down on the app icon or feature icon you want to free up space with (such as your photos) and then choose to free up space from the shortcut action.

These Google Pixel and Pixel XL hidden features make the phone even more useful. Now that you know them you can use them to get the most out of your new Google device.

How to Use QR Code Reader on Your Pixel 5 Phone

QR codes are supposed to make life easier, but having to install potentially shady third-party apps just to scan one is more trouble than it's worth. Thankfully, there's a QR code reader built into all Google Pixels, but you wouldn't know it unless you stumbled across the feature.

Scanning QR Codes

Open the default Camera app. Direct the camera app towards a QR code. You will see a small popup just above the shutter button with a preview of the QR code results. Tap on the popup, in our case it's a Wiki link. When you do, you will be directed to the specific location of the QR code, be it a website or a deep link within an app.

This also works for Wi-Fi QR codes. When the SSID, password, and security type are embedded into the QR code (for instance, when someone shares their WI-FI using Android 10's new QR system, your phone will automatically log into the network as soon as you tap the popup in your viewfinder!

Free Up Space through a Quick Shortcut

Some apps come with shortcuts. When you long press their icons from the home screen, you'll get a list of quick actions for that app. Long press the Twitter icon, for example, and you have the option to search, write a new tweet, or send a message.

Google Photos includes one of the most useful shortcuts, though. Long press the Photos icon, and you'll see a shortcut to "free up space." Hit that option and the app will find duplicate photos from your phone that are already saved on Google Photos. It will ask if you'd like to delete those photos since they've already been backed up.

Obviously, you can do this from the app itself, but this is a quick way to get there when you need some extra phone storage.

Customize Your Daily Briefing

"My Day" is one of the most fun, useful features in Google Assistant. Launch the app with the voice command, "OK Google, tell me about my day," and you'll get a briefing of the weather, traffic, news, and more. Best of all, you can customize what kind of information you want to include in the briefing.

From within Assistant, navigate to **Settings > My Day** and you can pick and choose what info you want: Weather, commute, meetings, and reminders. You can also customize your news sources by hitting the settings icon under News.

Get New Wallpaper Every Day

Pixel's live wallpapers not only look beautiful, some of them serve a practical purpose. Its horizon wallpaper, for example, shows a sunrise as your phone charges. When the battery fades, the sun goes down. You can also see a view of earth with real-time clouds, based on your location. (If you don't have the Pixel, Google has an app for them for other Android devices)

Google also includes a daily wallpapers feature that swap out your wallpaper image every 24 hours. To access these, head to Settings > Display > Wallpapers. If you choose Cityscapes, for example, you'll see an option to turn on a daily wallpaper. Tap the box to set it up.

Turn on the Notification Light

Pixel's notification light is turned off by default so you might not even know it's there. To turn it on, head to **Settings > Notifications** then hit the gear icon and turn on "pulse notification light." Whenever you get a new notification, a light will now blink periodically next to the earpiece speaker.

The light is subtle and it doesn't blink often, though. You may want to turn it on as a gentle reminder that you have notifications waiting for you, especially if you have your phone set to silent or vibrate all other times.

Play Trivia with Assistant

Okay, this isn't the most useful feature, but hey, it's kind of fun. If you ask Google Assistant to "play Lucky Trivia," it'll launch a quick game of trivia with five questions. You can play alone or with multiple contestants. It asks a series of random questions like the one above, and you respond by voice. The questions are pretty easy, but if you have kids, this might be a fun feature for them.

Access Split Screen Mode

For apps that support it (and even on many that don'ts), you can use split screen on the Pixel phone, too. To access split screen mode, open the apps you want to split. From within one of the apps, hold down the square menu button and your phone's screen will split in half. From there, you can choose which second app you want in the bottom half of your screen.

When you're in split screen mode, the square button aptly turns into two rectangle icons. To swap out the bottom screen app, hit this split screen icon. To exit split screen, hold down on the icon.

Automatically (and Securely) Connect to Open Wi-Fi Networks

Google's Wi-Fi assistant comes in handy when you want to save on data. It automatically connects to open Wi-Fi networks when they're available. It only connects to high-quality networks and secures your data via a VPN. When you're connected, a key icon appears in your status bar.

This feature is disabled by default. To turn it on, head to **Settings > Wi-Fi,** then hit the gear icon and turn on "use open Wi-Fi automatically."

Enable Night Light

Less blue light before bed is a good thing. Pixel's *Night Light* feature turns your screen tint from blue to red at night, kind of like f lux can but without the need to root your phone or use an app. Theoretically, less blue night at light helps you sleep better

Access Night Light from **Settings > Display > Night Light.** You can set it to turn off and on automatically from sunset to sunrise, or you can schedule it for specific times of the day.

Get Information Related to what's currently on Screen

Google Assistant will tell you all the interesting things it can do when you have the app open, much like Google Now on Tap used to on older devices. You can also use it from within other apps, though. Depending on what's on your screen, Assistant can offer extra information. If you're reading an article, for example, launch Assistant by saying "OK Google," and you can swipe up on Assistant to see suggestions for further reading or more information on a specific topic.

Pixel already comes with a surplus of practical settings, so it's easy to glaze over some of the more obscure features it comes with.

Use the Pixel's spam-blocking tools

We all get spam calls and texts. Fortunately, the Google Pixel makes them a lot easier to deal with — if you know what tools to use. When you get a spam call, for instance, you can fight back. Just head to the Phone app and press the clock icon to bring up your call history. Then, press and hold the spam number. You'll be able to select "Block/report spam," which will not only keep that number from calling you again, but will also report it to Google as spam.

Any phone number you block will be prevented from texting you. And the list of numbers that you block will be synced to any Android phone that you upgrade to in the future. You can manually make additions to the list of blocked numbers by tapping the settings icon in the upper right corner of the Phone app, tapping "Settings," and then hitting "Call blocking." And while you're in the Phone settings, Raphael recommends looking at the "Caller ID & spam" option to make sure it's activated. That way, you'll see the name of every incoming caller, even when it isn't in your address book. And you'll be notified when the number has been reported as spam.

Capture the highest-resolution photos and videos

One of the Google Pixel's best claims to fame is its record-breaking camera. But if you want to make the most of that camera (and of the free, unlimited photo and video storage that Google offers for Pixel users), you'll want to make sure you're always capturing photos and videos at the highest resolution. To do that, you'll just need to change a few of the default settings in the Camera app.

Open the Camera app and tap the menu icon in the top left corner. Select "Settings," and tap on the option for "Back camera video resolution." Make sure it's set to UHD 4K. Also look for the option for "Panorama resolution," and make sure it's set to "High." Since you have free photo and video storage, it makes sense to use the highest resolution. It will really demonstrate the extent of the Pixel's photographic prowess.

Free up space by quickly deleting duplicate photos

Android apps come with shortcuts_you can access by long-pressing their icons from the home screen. But it's not just third-party apps that come with this useful functionality. In fact, Google Photos has what Wong characterizes as "one of the most useful shortcuts." (Especially true when you're always looking for ways to free up storage on your phone.)

Long-press the Photos icon and you'll get a shortcut that promises to "free up space." If you hit that option, then the Google Pixel will go ahead and find duplicate photos on your phone that are already saved on Google Photos. It will then ask you if you'd like to delete those photos from your phone's local storage, since they've already been backed up in the cloud. Wong notes you can definitely accomplish that task on your own. But the shortcut is a fast and easy way to get the functionality when you need a quick way to free up some storage.

Take a photo without hitting any buttons

Another photo-related trick every Pixel owner needs to know? You can actually take a photo with either the front-facing or the rear-facing camera without having to press any buttons. To do that, you'll just need to familiarize yourself with Google Assistant and a few of the new voice commands you can use to get things done.

Using Google Assistant to take a photo is actually pretty easy. And the voice commands you need to accomplish the task are simple to remember. Just say, "OK Google, take a picture" to start a three-second countdown to a capture with the rear-facing camera. Or say, "OK Google, take a selfie" to get a three-second countdown until the front-facing camera captures a snapshot. You can even say, "OK Google, take a picture in fifteen seconds" if you need more time.

Free Cloud Storage and Smart Storage

Google furnishes the Pixel with free and unlimited cloud storage for photo backups. Combined with Smart Storage, you can take an unlimited number of photographs without running out of space. It works like this: based on how you've configured your Pixel, your phone might delete images every 30, 60, or 90 days — but only if your photos have been backed up. Plugging your phone into a power source triggers the backup process.

Configuring Smart Storage

To configure Smart Storage, navigate to **Settings** and the choose **Storage**. In the Storage menu, you can toggle both the backup frequency and whether or not a backup occurs. Unfortunately, Google does not integrate (nor does it permit integration) with a third-party cloud storage solution like OneDrive. Because Smart Storage links into Google Photos that means you get unlimited photo backups___unlike other smartphones, the Pixel comes with full resolution quality. The unlimited storage that comes with other services usually means the photos are compressed and of diminished quality.

Facial Recognition

One of the creepiest features in Google Pixel is its photo app. Once you snap a picture of friends, Google uploads your photos to the cloud, and its AI applies a facial recognition algorithm. Each and every person in your album can receive a tag (or name). You can search your database for all pictures of a specified individual. But that also means Google keeps a catalog of not just your identity, but the identities of your friends (and random strangers who blundered into a shot). The potential consequences remain unknown — but the risk of abuse chills even the staunchest of technology advocates.

Concierge Service

If getting access to a machine assistant wasn't enough, Google also borrowed a feature from the Amazon Fire Phone: direct access to a human assistant. If your Pixel ever malfunctions — or you just need some advice — a *human* operator can help with just the touch of a button.

The Hidden RGB LED Light

Among its unadvertised features, the Pixel throws in a secret LED for notifications. Users can enable this feature, which saves on both battery life and screen lifespan. To permit it, the user must navigate to **Settings > Notifications** and click on the **gear icon** in the upper-right side of the screen. Then enable the **Pulse notification light**. From then on, whenever users receive a notification, the LED blinks.

Twist-for-Selfie

Like the Moto X, the Pixel can switch between cameras with ease. It works like this: after opening the Camera app, make a double-twist motion while holding the Pixel to launch the selfie camera. The Pixel then switches from the rear-facing camera to the front-facing camera.

Lift-to-Check-Phone and Double-Tap to Wake

Ambient Display turns the screen on temporarily to show notifications. On an AMOLED panel, this saves battery life and helps with screen burn-in.

Newer Android devices usually include Ambient Display by default. However, the Pixel adds two additional features. Lift-to-check-phone allows the Pixel to switch its screen on whenever it senses that someone has picked the phone up. Double-tap to wake spares the physical power button by waking the device by double-tapping on the screen.

Thank you for purchasing this guide and I believed you learned some tips and tricks to maximize your device.

Made in the USA
Monee, IL
28 May 2021

69664473R00105

Are you looking for a manual that will expose you to all the amazing features of your device? Then get your hands on this book and have an amazing time using your device.

Google released the fifth iteration of its flagship smartphone, the Pixel 5, which embodies the tech giant's vision for what a smartphone should be. This manual will teach you everything you need to know to master your device

ISBN 9798694851916

90000

9 798694 851916